Book 3, A LITTLE BIT OF DINOSAUR SERIES

# A LITTLE BIT OF THAT DINOSAUR!

by Elleen Hutcheson and Darcy Pattison

Illustrated by John Joven

A Little Bit of THAT Dinosaur!
by Elleen Hutcheson and Darcy Pattison
illustrated by John Joven

Mims House
1309 Broadway
Little Rock, AR 72202
USA

MimsHouseBooks.com

Publisher's Cataloging-in-Publication Data

Names: Hutcheson, Elleen, author. | Pattison, Darcy, author. | Joven, John, illustrator.
Title: A little bit of THAT dinosaur! / By Elleen Hutcheson and Darcy Pattison; illustrated by John Joven.
Description: Little Rock, AR: Mims House, 2023. | Summary: This humorous story follows a nitrogen atom as it journeys from a dinosaur egg to your thick skull.
Identifiers: LCCN 2022919219 | ISBN 9781629442280 (hardcover) | 9781629442297 (paperback) | 9781629442303 (ebook) | 9781629442310 (audio)
Subjects: LCSH Life cycles (Biology)--Juvenile literature. | Nitrogen in the body--Juvenile literature. | Bones--Juvenile literature. | Dinosaurs--Juvenile literature. | Humorous stories. | CYAC Life cycles (Biology). | Nitrogen in the body. | Bones. | Dinosaurs. | BISAC JUVENILE NONFICTION / Science & Nature / Biology | JUVENILE NONFICTION / Science & Nature / Anatomy & Physiology | JUVENILE NONFICTION / Science & Nature / Fossils | JUVENILE NONFICTION / Animals / Dinosaurs & Prehistoric Creatures
Classification: LCC QH501.H88 2023 | DDC 372.3/57--dc23

You have a little bit of Hadrosaurus in your thick skull!

Don't believe me?
It's all your cousin's fault.
Listen up.
Here's how
it happened.

Once, in days of old, just like now, the air was full of nitrogen.

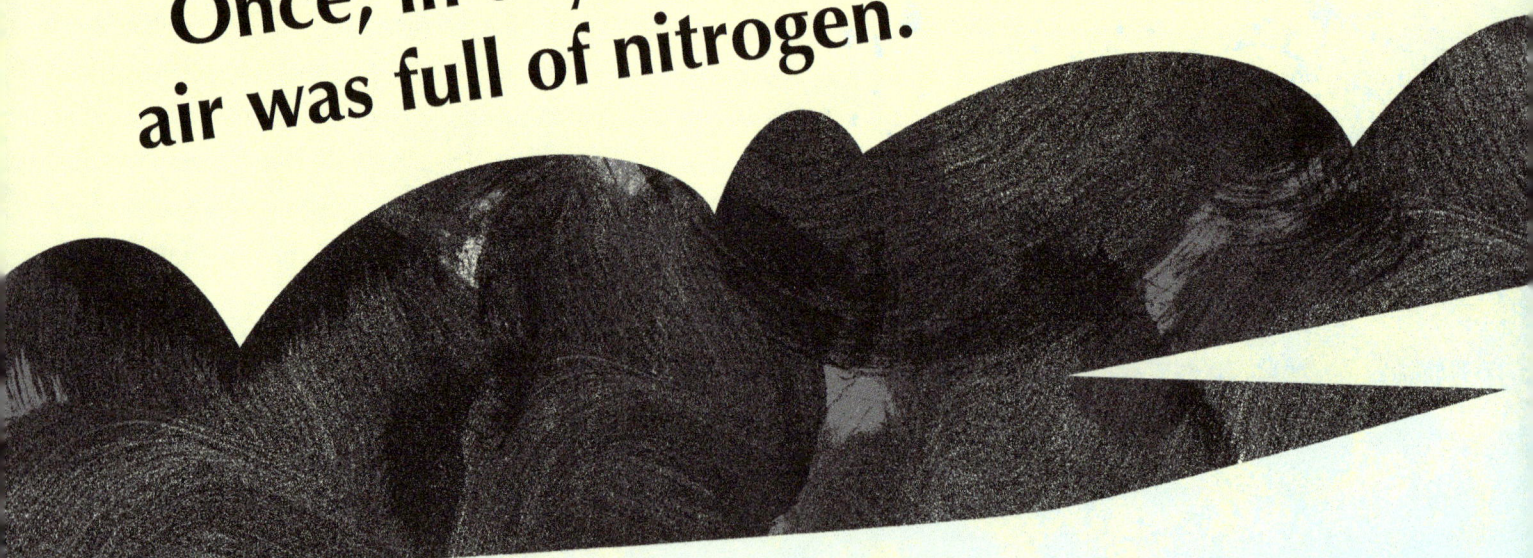

Mostly it was two nitrogen atoms stuck together.

A storm rolled and brewed.

NITROGEN ATOMS!

Lightning flashed! The heat from the lightning split the two nitrogen atoms apart.

One nitrogen atom grabbed an oxygen atom to become nitrate.

NITRATE

The nitrate fell with the rain and soaked into the soil. There, it was taken up by a horsetail plant. A hungry Hadrosaurus lumbered along.

She was so big that she was always hungry. When she saw the horsetail plants, she gobbled them up, and the nitrogen atom became a little bit of the dinosaur.

The mother Hadrosaurus migrated to nesting grounds and built a nest. When she laid an egg, the nitrogen atom became part of the egg's shell.

But a landslide covered the egg with mud.

The egg hardened into rock.

Time ticked by. The land became fields.

A farmer planted peanuts, which grew and sent out roots. Bit by bit, water washed away a little bit of dinosaur from the fossilized Hadrosaurus egg.

Bacteria in the peanut roots absorbed the nitrate and changed it to usable nitrogen so the plant could make peanuts.

The farmer harvested the peanut with the nitrogen atom that was once a little bit of dinosaur.

The peanuts were packaged, and a trucker delivered the peanuts to the ball field.

You sat with your cousin and watched the baseball game.

You both cracked peanuts and ate them. Your cousin DARED you to throw a peanut into the air and catch it with your mouth.

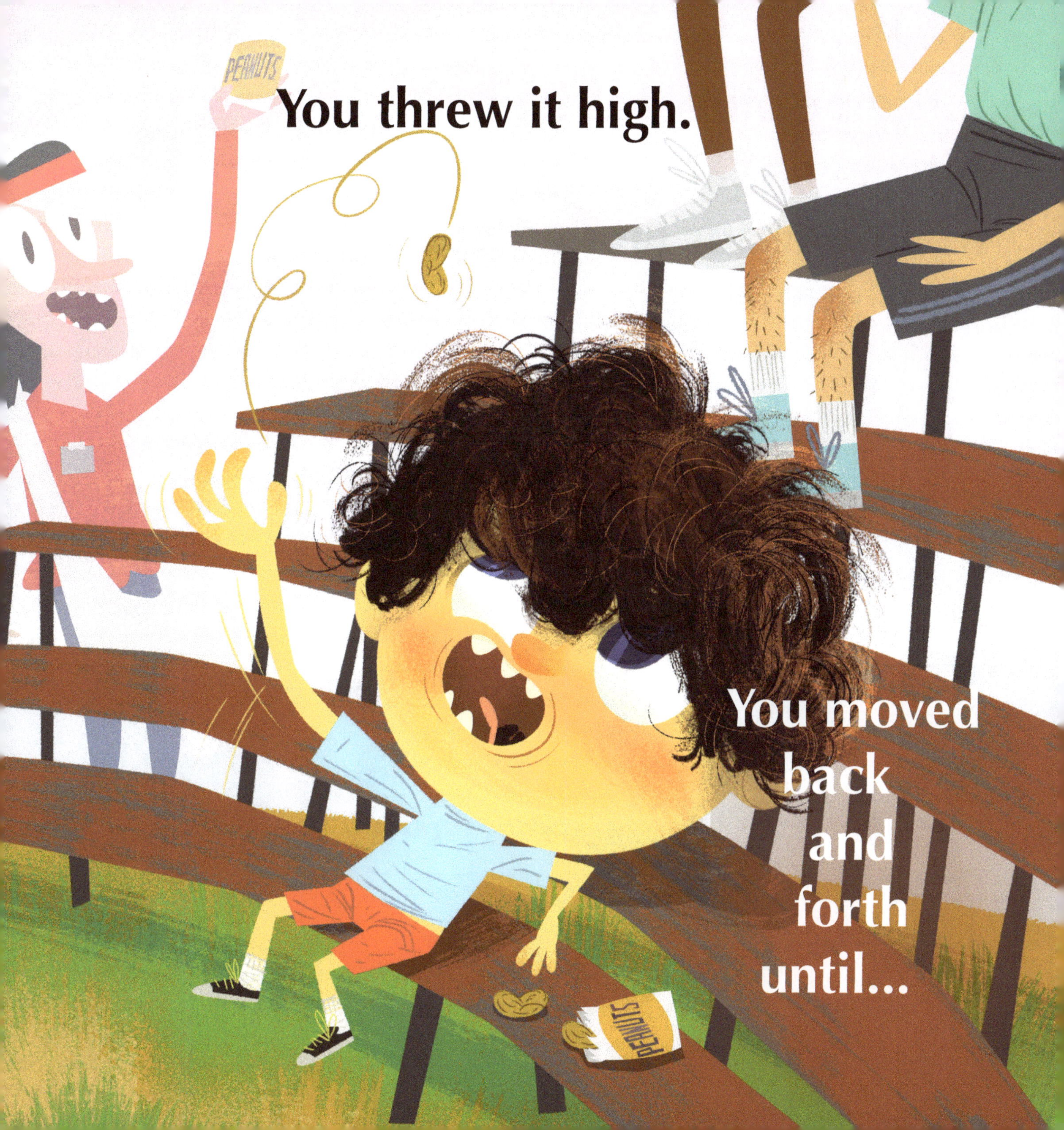


You threw it high.

You moved back and forth until...

...the peanut that had nitrogen that used to be in a dinosaur landed in your mouth.

(I told you it was your cousin's fault.)

Your body took that little bit of dinosaur and used it to make your body a little bit stronger.

You have a little bit of Hadrosaurus in your thick skull.

If you look closely, you might see a little bit of archaeopteryx in your feathered face,
OR . . .

...a little bit of ankylosaur in your knobby tail, OR...

...a little bit of iguanadon in your strange hands, OR...

...a little bit of gigantosaurus in your thick legs. And when...

...your days on earth are done, and your body returns to the land, that little bit of dinosaur will be used again.

**Maybe someday the nitrogen will**

# travel

**halfway around the world and wind up in the beak of a little blue penguin.**

# The Nitrogen Cycle

All matter and all living things are made up of atoms, tiny particles that you can only see with powerful microscopes. The main atoms that make up living thing are carbon, hydrogen, oxygen, nitrogen, phosphorous, and sulfur. When any living thing dies, the atoms that make it up are used over and over in new living things. The atoms go from a living thing to the soil, air, or water, and then back into a living thing.

Nitrogen in the air starts the nitrogen cycle. Two nitrogen atoms are stuck together until lightning makes them split. One nitrogen atom combines with oxygen to make nitrate. Plants use the nitrate to build leaves, roots, or fruit. When animals eat the plants, the atom becomes part of the animal, and maybe part of an egg it lays.

If the egg is fossilized, then the nitrate stays in the ground a long time. Eventually, water leaches away the nitrate and it's absorbed by another plant. If an animal or person eats that plant, then the nitrogen becomes part of them. When they die and are buried, another cycle starts all over again. Another plant can use the nitrogen atom, then another animal can eat the plant, and the nitrogen becomes part of that animal. The cycle can repeat endlessly.

www.ingramcontent.com/pod-product-compliance
Lightning Source LLC
Chambersburg PA
CBHW052346210326
41597CB00037B/6278